Walter Porter Manton

Taxidermy Without a Teacher

Comprising a complete manual of instruction for preparing and preserving

birds, animals and fishes. Second Edition

Walter Porter Manton

Taxidermy Without a Teacher
Comprising a complete manual of instruction for preparing and preserving birds, animals and fishes. Second Edition

ISBN/EAN: 9783337241230

Printed in Europe, USA, Canada, Australia, Japan

Cover: Foto ©berggeist007 / pixelio.de

More available books at **www.hansebooks.com**

TAXIDERMY WITHOUT A TEACHER

COMPRISING

A COMPLETE MANUAL OF INSTRUCTION

FOR PREPARING AND PRESERVING

BIRDS, ANIMALS AND FISHES

WITH

A CHAPTER ON HUNTING AND HYGIENE; INSTRUCTIONS
FOR PRESERVING EGGS AND MAKING SKELETONS
AND A NUMBER OF VALUABLE RECEIPTS

BY

WALTER P. MANTON

Illustrated

SECOND EDITION, REVISED AND ENLARGED

BOSTON
LEE AND SHEPARD PUBLISHERS
NEW YORK CHARLES T. DILLINGHAM

CONTENTS.

Fig. 1.

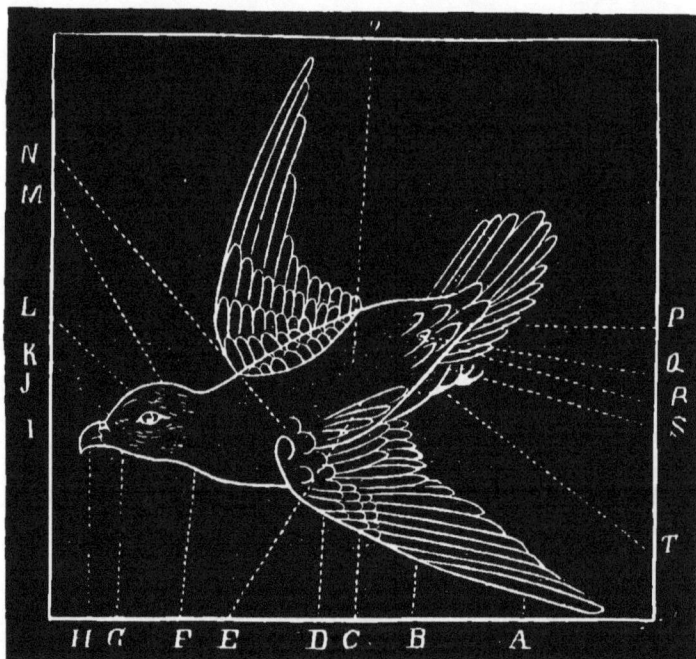

A — Primary Quills.
B — Secondary Quills.
C — Spurious Wing.
D — Wing Coverts.
E — Tertiary Quills.
F — Throat.
G — Jugulum.
H — Beak — Upper and Lower Mandible.
I — Culmen of Upper Mandible.
J — Cere.
K — Commissure.
L — Frons, or Forehead.
M — Occipital Feathers.
N — Scapular Feathers.
O — Back.
P — Upper Tail Coverts.
Q — Rump.
R — Shows position of Under Tail Coverts.
S — Abdomen.
T — Tarsus.

PREFACE.

THE success of this little book during the past six years necessitates a second edition. As a manual it is not intended to compete with the larger handbooks on the subject; but the attempt has been made to furnish the beginner with reliable instruction for the least money possible. The present edition has been thoroughly revised, and many additions made.

As was said in the first edition : " I have employed the method given for a number of years, and with great success, and guarantee success to the learner who fully carries out the directions embodied herein. I ask the reader to take himself, in imagination, to my work-shop, and to proceed as if I were at his elbow, guiding his hand, and explaining to him the mysteries of this beautiful art. It is only continuous, untiring labor that accomplishes anything of real merit in this life; and the most successful ornithologists will be found to be the hardest workers.

" Therefore I would caution the beginner against all impatience and disappointment at unsuccessful attempts, and urge him to press forward, continually striving to improve upon past failures, and soon, to his own astonishment, those things which at first appeared difficult and awkward, will become comparatively simple and easy. Said an old teacher to me : ' I can tell you how all these things are *done*, but I cannot enable you to do them ; practice alone will accomplish that.'

" A person with a light and delicate touch will be most successful in this art ; therefore I recommend it to the special attention of ladies. It is a continual source of pleasure, and promotive to the love of the great Nature which moves so mysteriously around us. It is true that we have seen those of coarse and vulgar minds and clumsy fingers, eminently successful ; but what is more revolting to a delicate appreciation, than to see these bright creatures, so marvellously constructed by our all-wise Father, tortured into life-like attitudes by one who acts merely as an automaton, and has no sympathy with his work otherwise than to gain a livelihood? It is only the refined and the lover of nature who can thoroughly enjoy this art of reproduction. A close observer of nature, in two short hours spent in the fields and woods, will see and learn more than the unobserving and careless person in as many years.

"A careful observation of the habits and attitudes of the little songsters when free, will be of great assistance in mounting. A knowledge of drawing will also be found of service."

FROM THE FIRST EDITION.

Brown University, Providence, R. I.

Mr. Manton, — Having perused your MS. "Taxidermy Without a Teacher," I feel free to say, that its suggestions are eminently practical, and cannot fail to render such aid to the beginner as he most needs, and indeed must have from some source, at the outset of his efforts to acquire the beautiful art of preserving and mounting specimens in Natural History.

Whatever induces the young or old to turn their attention to the study of nature, is a gain to society at large, as substituting truth for fiction, and leading the mind to the contemplation of Him whose devising wisdom and sagacity are manifested in all His works. Commending highly your effort, I am yours,

With great respect,

J. W. P. Jenks.

HANDBOOK OF TAXIDERMY.

CHAPTER I.

BIRD SKINNING AND MOUNTING.

WELL, here we are at last. Please turn the key in that door — to keep all inquisitive priers out — for the process into which I am about to initiate you is something of a secret, shrouded by the thin veil of mystery.

You have come to me to-day to learn something of the art of Taxidermy, so we will take up, for your first lesson, bird skinning and mounting. But first let us see what

TOOLS

we shall need to accomplish our end: a pair of good sharp scissors — surgical scissors, with long handles and short, stout blades are the best; a knife or scalpel; a pair of spring forceps; a common knitting-needle; a rabbit's foot, which should be cut off at the knee, the nails cut out, and thoroughly cleansed and dried, — used for smoothing and dusting the feathers of birds after mounting; a fishing-hook, with stout cord attached, for suspending the bodies of birds that would otherwise be too large to handle conveniently.

On the whole, I would advise you to get at the start a common dissecting-case, which will contain all of the above, and besides being convenient, may save you much delay and vexation.

You will need a pair of stout wire cutters; a flat file; a pair of wire twisters or forceps; plenty of pins; thread and needles, — surgical or saddlers' needles, as they are called, are the best, as they cut instead of punching the skin; a brain scoop, made by twisting a bit of wire into a loop; and a bobbin of thread, which you can procure at any cotton factory. You should also have on hand an assortment of annealed wire; glass eyes of various sizes and colors; tack nails; brads; a piece of putty; sealing and bees-wax; paints; glue; artificial leaves; mosses; everlasting flowers, etc., for ornamenting perches.

Now we come to the

required for stuffing. Cotton, tow, soft hay and excelsior are the best, but anything soft will do, except feathers, hair, or in fact any animal substance, as they act merely as assistants to the taxidermist's great enemies — the bugs (Tineidæ and Dermestidæ).

POISON

is used to preserve the skins. But as everything of that description is dangerous for young and inexperienced persons to handle, I strongly recommend the following preparation, and guarantee it to preserve their first efforts until they become an eye-sore, and are finally thrown into the fire with much disgust.

I.

Pulverized Alum.

Common Salt. Equal parts.

Mix.

Label: Salt and Alum.

The best and only *safe* preparation is : —

II.

Arsenic.

Pulverized Alum.   Equal parts.

Mix.

Label: POISON ! !

The arsenic is to poison, and the alum to act as an astringent, especially in setting the feathers and fur of skins partially decayed. As arsenic is an irritant poison, great care should be taken while using. See that the hands are free from all scratches, cuts, hang-nails, and broken skin. These may be covered with court-plaster or collodium. Wash the hands immediately after using, and be careful to clean well under the nails. With these precautions there is little or no danger, and it may be used with the greatest impunity. Avoid all so-called "arsenical soaps," as they are both dangerous and disagreeable to handle. Use nothing but the above receipts, and you will succeed far better. Having all these materials and implements at hand, we are now prepared to go on with our work.

LABELLING.

Let us take this Blue Jay for your first attempt. The first thing to be done is to measure
and label it — and, by the way, never neglect
this, for a bird without its label in a collection,
is like a ship at sea without its rudder. LENGTH
— Lay the bird on its back, and with a pair of
dividers (for a large bird a tape line must be
used) measure from the tip of the beak (the
head lying flat on the table) to the tip of the tail.
Place the points of the dividers on a rule that is
divided into one-hundredths of an inch, and
see how much they measure. EXTENT. — Place
the bird across the ruler, and using reasonable
force, stretch the wings out, and see how far
they reach. LENGTH OF TAIL. — Place one
point of the dividers at the end of the "pope's
nose," and open them until the other is at the
tip of the longest tail feather. THE TARSUS. —
Place one point of the dividers at the middle of
the sole of the foot, and measure as far as the
first joint. THE BEAK. — Place one point of the
dividers at the beginning of the cere, on the

upper mandible, and open them until the other is at the tip of the beak. In addition to these I advise you to keep the weight of each specimen, especially in the case of game birds. Set all these measurements, etc., down on your label as you go along; also color of eye, contents of stomach (after skinning), and the number of the bird. This number must correspond to a number in your Ornithological Ledger — a book in which you should keep an account of each day's doings; the number of birds killed, the number used, attitudes, etc., and whatever else may be of interest to you regarding the day's shooting.

BLOOD STAINS.

These may be removed before skinning, by gently washing with a sponge and a little water, and afterwards dried by working into the feathers pulverized plaster of Paris, or potato starch, until the water is all absorbed, and the feathers become dry and clean; then shake all plaster or starch from the feathers. Now fill the beak, anus, and shot holes, if you have not previously done so,* with cotton, and we are ready to begin

SKINNING.

Lay the bird on its back, its head towards your right hand, and run the handle of your scalpel from the sternum, or breast bone, to the anus. In so doing you will see there is a little naked place, in many birds, all the way down. Stroke the feathers away right and left, leaving this bare, and inserting the point of the scissors at the end of the sternum, cut down to and *into* the anus (taking care not to cut through the thin belly walls; if this is done, fill the place with cotton, or disembowel); stopping here, as this makes a good strong termination that will not easily tear. Take the forceps in the right hand, and seize one edge of the skin. Holding this, press and push (never pull) the skin from the sides and belly walls. Care must be taken that the feathers do not get into the cut and thus become soiled. Keep stroking them away, right and left, and place a little fluff of cotton, tissue paper, or white pine sawdust, under them. After

---

* See Hunting and Hygiene.

skinning away, you will come to a hard substance; this is the thigh. Skin carefully around
this until you come to the under side, when you
can easily insert your scissors and sever it from
the body. Push the leg up out of the skin until
you come to the tarsus; clear away all muscles
and tendons, and bring the legs back into its
skin again. Repeat this process on the other
side without turning the bird around. Now
skin carefully around the tail; place your forefinger across this, and pressing it back a little,
insert the scissors and sever the stump. Great
care must be taken, however, not to cut the thin
and very tender skin over the tail.

Now turn the bird up, and with its belly pointing toward you, let the tail fall over the forefinger of your right hand, and with your thumb
nail and fingers, continue to push and work the
skin until you come to the wings; sever these at
the shoulder.

Now holding the skin in the left hand, and
letting the body fall over the other side of the
fingers, skin down the neck — which will slip
out as easily as a finger from a glove — until you

come to the base of the skull. Skin carefully over this, taking great care to detach the thin membrane of the ear, with the thumb-nail or scalpel handle, and proceed until you come to the front part of the eye socket. Cut the thin membrane that covers the eye, taking care not to lacerate the ball; then scoop out the eyes. Stick one point of the scissors just inside one branch of the lower jaw, and make a cut parallel with the jaw, crushing through the skull just outside the angle of the jaw. Make a duplicate cut on the other side. Then at the end of these make a transverse cut through the roof of the mouth.

Connect the posterior ends of the side cuts by cutting across the skull near its base. You have now cut out a square-shaped piece of bone and muscle, and by pulling gently on the neck, this will come out, bringing with it a mass of brain. Remove all brain and muscles of the head. Skin down the wings as far as they will go, and run the thumb-nail along the ulna, detaching the quills to the metacarpal bones; remove all muscles and tendons. Now turn the skin and shovel in arsenic, so that all parts may be covered; afterwards shake the skin over your box to remove all loose arsenic.

Some difficulty may be experienced in getting the head back into the skin. Begin in any way you please until you see the point of the beak coming through the feathers; seize this with the fingers, and making a cylinder of your left hand, gently coax the skin backwards, with a motion very much like that of milking.

Now if you wish to make the skin neat, dress every feather with the thumb and knitting-needle, and see that they all lie in place. Insert the knitting-needle through the eye to the top of the skull (under the skin), adjust the scalp and see that every feather is smooth.

In birds with large heads — such as owls, some woodpeckers and ducks — over which the neck skin will not easily slip, a slit must be made along the top of the head and the skull worked through, and treated as given. When completed, sew up the skin and carefully arrange the feathers.

When birds are to be mounted with spread wings, as if flying, it is sometimes desirable to make the incision along the back **instead** of the belly, the ventral feathers thus **presenting a** smoother appearance.

MAKING A SKIN.

After a skin has been poisoned and dressed, it may be "made" by inserting into each eye-socket, through the neck, with the knitting-needle, a little ball of cotton. Then make a little roll of cotton and insert it into the neck; one end in the cavity of the skull, the other just appearing at the end of the neck. Some collectors at this point fasten the wings to the sides, by taking a stitch through them with needle and thread. Before doing this, be sure that the wings are in the right place. Take a piece of cotton about one-half the size of the bird's body, and by turning in the edges make it into an oblong ball, corresponding to the body just removed. Place this in the skin with the forceps, and before letting go with the thumb and forefinger press the wings together on the back, placing the fingers *under* the wings. Now draw the edges of the skin together, and making a cylinder of each hand, gently coax the skin through, until it is of the required shape. Then place it in a drying-rack, made by bending a piece of zinc or tin into a half cylinder. Leave it to dry for a few days. Many collectors never mount birds, but prefer "made skins." These may be relaxed at any time by wrapping in damp cotton for a few days, and then set up as directed.

SEX

May be determined by cutting through the ribs under the right wing, and pushing away the intestines. There, bound to the small of the back, will be seen the *testicles* of the male — two spheroidal, whitish bodies, which vary in size according to the season of the year. In the female will be seen the *ovaries*, a flattened mass of whitish bodies. These are often so minute as to defy the naked eye, and the inquirer is obliged to employ the microscope to make the distinction. The sign recognized by ornithologists all over the world is ♂ for males, and ♀ for females; to which is added for young birds the Latin *juvenis* or *juv.* or O, meaning young, and *Nupt.* for birds in nuptial or breeding plumage.

In the first place we must prepare the wires that we shall need. There are three of these —

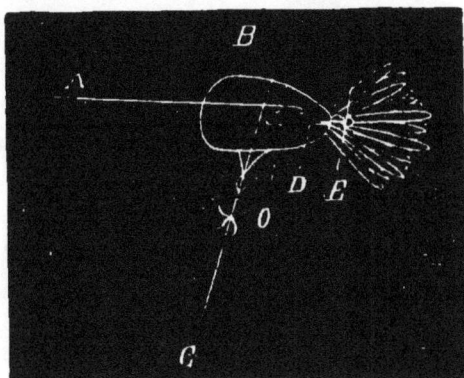

FIG. 2.

the head wire and two leg wires. The first of these must be about three or four inches longer than the bird as it lies stretched out on the table (Fig. 2, A); the second and third two or three inches longer than the leg (C).

These wires must be perfectly straight (in the case of small wires this may be done by stretching), and have one end sharpened. To do this make a little groove with the file in the table, lay the end of the wire in this, and holding it in the left hand, place the forefinger near the end, and seizing the file in the right hand slowly sharpen, revolving the wire at the same time with the left. This will seem a little awkward at first, but you will soon become accustomed to it. Now take a piece of sand-paper and polish the wires.

Take the longest of the three wires, and bend the unsharpened end into an oblong ring — in length according to the size of the bird to be mounted. Or, instead of the ring, you may make a little oval block of wood, secure the head wire to one end, and bore holes for the leg wires, which must be firmly fastened. For this Blue Jay we will make it about one inch in length. Now around this ring or block as a *nucleus*, or foundation, place. the tow and wind it on with thread or string, continually putting on more tow until you have an egg-shaped form (B). Wind *around* and then *lengthways* to accomplish this. For birds larger than a canary, the body may be made of hay or excelsior, and finished with a coating of tow. This is easier to put wires through, and is more economical. The tow body must be as near the size of the natural body as possible, if anything a trifle *smaller*, on no account *larger*. In order to be more accurate, I generally keep the body of the bird skinned, on my table, and while winding.compare the artificial body with this until it is perfected. In this way a better shaped and firmer body is produced. Be careful that you do not get the body too soft (you cannot get it *too hard*), or when you come to set up your bird it will be too weak to stand on its legs; the wires will have nothing to clinch and hold to. Now take a bit of cotton, and with the forceps introduce into the eye-socket through the neck. Repeat on the other side. Fill up the cavity between the mandibles and the space in the cranium with finely cut tow. Unless you are making a " skin," this had better be done directly after poisoning the skin, before turning the head through the neck. Now as to the neck. Some say, " Fill out gently with chopped tow." I prefer to wind the wire, A, for a short distance, with a bit of tow. To make this stick, first rub the wire with a piece of beeswax. This, I think, makes a better neck, and is less liable to misshape and contort the skin.

Now push the leg wire through the sole of the foot, and run it along the leg-bone up through the leg. Great care must be taken not to break the tarsus or run the wire through the loose skin which envelops the leg so as to tear it. Repeat on the other leg. Now wind a little splint of tow around the bone and wire of each leg. This will require some practice, but once acquired it is very easy. Place the body in the skin, and with a twisting motion run the wire out through the top of the head. Gently draw the skin over the body until it is about half way in. Then run the leg wires through the body, a little front of the middle and a trifle higher up. When the wire appears through the other side, seize it with the twisters, and bending it into a hook, draw it firmly into the body. Repeat on the other side. Now work the body entirely into the skin, by bending lengthwise the legs, and gently sliding them on the wires. This done, take a little chopped tow or cotton and place it under the body, on each side, directly where the shoulders come. Pin or sew the edges of the skin together. There will be a long piece of wire projecting from the head. Cut this off, say quarter of an inch from the head ; and making a ring at the unsharpened end, push it through the stump of the tail into the body. This is to support the tail (D).

Just now the bird is a decidedly shabby looking affair, and if you are not careful you will find yourself getting discouraged, and thinking that you have spoiled the skin. But do not despair, for if you have carefully followed the directions, all will be right, and you will soon have the pleasure of seeing a well-mounted specimen. But it will only be after many failures that you will succeed. Remember " Post nubila Phœbus," — and it is just the same with bird stuffing.

The legs are now straddling wide apart. Bring them together parallel to each other, and make a sharp bend at the knee, bringing them over the body in a natural position. Now place the bird on a temporary perch; bend back the head, and arrange the body in the position you think most lifelike. Some difficulty may be

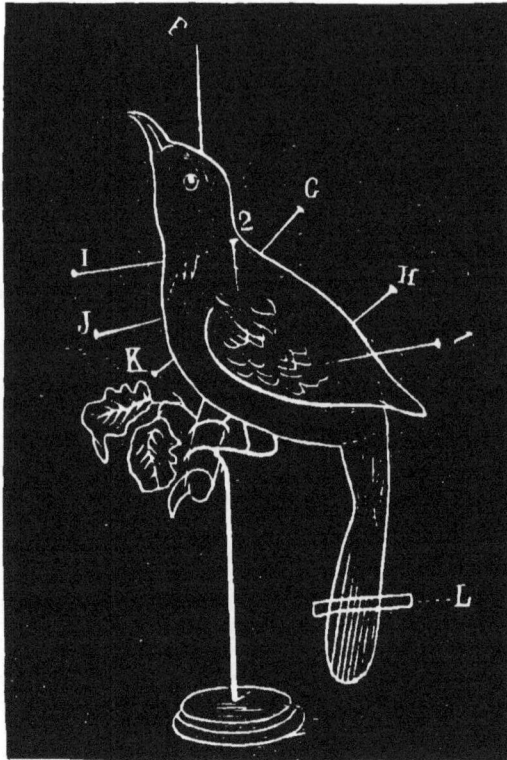

FIG. 3.

found in adjusting the wings, but if preceding processes have been rightly carried out, they will readily fall, or may be easily worked into place. Stick two pins through them at right and obtuse angles to hold them in place (1 and 2, Fig. 3). Now your bird will look much better, and with the exception of rumpled feathers, quite lifelike. To remedy this latter, with the knitting-needle and thumb go all over the bird with a kind of picking process; lifting the feathers and letting them gently fall into place. You cannot work over the bird too long in this way; and the more time you spend in dressing the feathers, the better will be your specimen after drying. Now stick two pins along the back, and three along the breast (G, H, I, J, K). Fasten one end of the thread from the bobbin to the projecting head wire, and carefully wind the entire bird. Do this loosely, so as not to disarrange any of the feathers, tightening, however, wherever they tend to rise or look uneven.

This winding process is considered by some to be the most difficult part of bird mounting.

The specimen should now be set away for several days, or even weeks, if the bird be a large one, and allowed to dry; after which it may be unwound; the eyelids soaked, by inserting little flabs of wet cotton until they

become soft and pliable ; the eyes inserted in
putty, and the lids carefully adjusted over them.
The protruding head wire and the pins in the
wings are cut off, and your bird mounted on
the perch which you have already prepared
for it.

WINGS AND TAIL.

Wings may be spread by running a wire
through the primaries (Fig. 1, A) into the body,
and placing another near the end of the wing as
a support while drying. The tail may be spread
by running a wire through the quills, near the
"pope's nose" (Fig. 2, E), or by placing a bit of
split wood across the tip and tying the open end
firmly (Fig. 3, L). Crests may be raised or spread
by inserting a small fluff, or ball of cotton under
the feathers, using a pin to hold them in place.

When glass eyes are not at hand, black beads
may be used. Or white glass beads may have a
pupil (black) surrounded by the iris (yellow or
brown) painted with oil colors on the back. If
neither of the above can be procured, a half
globe of the right size may be cut out of cork or

wood and a pin run through its centre. The outside is then to be covered with sealing-wax or varnish until quite smooth, and then painted the required color.

The feet, tarsi, cere and loose skin about the necks of some birds often fade or become dull. These should be carefully painted, imitating the original colors as closely as possible.

You have now completed your first lesson, and I advise that you become perfectly familiar with skinning and mounting birds, before you take up that of animals. For you cannot become too familiar and too much at home in this department; and it will come in play fifty times, where the other does once.

CHAPTER II.

PERCHES.

A VERY pretty perch can be made by arrang-
ing wires in the shape of a twig or branch, having
one end firmly fastened in a block of wood.
Wind the wires to the proper size with tow,
and after giving the whole a coating of thin glue,
sprinkle over it smalts and dry moss, rubbed fine
in the hands; when this is dry, you can glue on
artificial leaves, flowers, and grasses as your taste
prompts.   Another good perch for small birds is
a stump made of pasteboard, with a small opening on one side.   Cover this with the same materials as above.   It should be about an inch and
a half in height.   Another perch is made by re-
ducing pasteboard to a pulp, and moulding it
around a twig or wire form.   Boil the pasteboard to a pulp in a little water.   Then force
through a coarse sieve, and mix with thin glue.

Mould this around the form, give it a coat of brown paint, and decorate to taste. To give it a rougher appearance, a coarse comb may be drawn over it before painting.

A very effective way of mounting humming-birds is to form a tree with small palm leaves, or others, attached to a moss-covered stalk, having moss and grasses at its base. Let the hummers have the wings and tails spread, and crests and breast-tufts raised in the most effective manner. They are then attached to single wires starting from the limbs of the tree, with back or breast showing according to the part which is to be displayed.

FIG. 4.

Another good way, especially where the collection is large, is a single wire bent oval, and both ends fastened to a block standard. To the outside and inside are fixed short perches of wire upon which the birds are mounted.

Birds mounted for ornamental purposes should be placed under glass, to protect them from dust and insects. A very neat homemade case is constructed of window-glass, cut of the proper dimensions; the sides, top and bottom being fastened together by strips of stout paper glued over their edges. Or the glass may be set in a light framework of wood, which may be painted, stained, or ebonized. To ebonize, you require extract of logwood, a supply of rusty nails, or scraps of iron, and some vinegar. Place the iron in the vinegar a week or more before using the latter. When you are ready to begin, give the wood several coatings with a strong solution of the logwood, and when this is nearly dry, brush over with the vinegar. A fine dull black color will be produced.

All ducks, wading and ground birds should be mounted on a piece of board; and long-legged birds should have one foot a little in advance of the other, as if in the act of stepping. The attitudes of birds, seen in your rambles, may be put to use in your mounted specimens, and your own taste will suggest a variety of perches and ways for mounting.

CHAPTER III.

SKINNING AND MOUNTING MAMMALS.

WHEN the beginner has once become proficient in skinning and mounting birds, he will have but little difficulty in "setting up" mammals. The same general principles are to be observed with each.

SKINNING.

Cut with the scalpel or stout scissors from the breast-bone down to the anus: sever the legs close to body, and treat both legs and head as given for birds.

Some difficulty may be experienced in skinning the tail. This is readily done if it be a hairy tail, by pushing the skin over the first two or three vertebræ, then seizing the stump with the left hand, pull, at the same time holding the skin back with the right hand. The bone will generally slip out as easily as a sword from its sheath; but if it will not come, tie a knot of strong cord over the end, and fasten to some support firmly. Then holding on with the right hand, as before, you can easily strip the tail to the tip.

MOUNTING.

Instead of three, you must now have five wires. Sharpen and sand-paper, as the former, and make a nucleus for body. The shape of an animal, with the neck severed from the skull, is like the italic f laid on its side ($\leftharpoondown$). This is made by winding the tow on the nucleus, the same as with birds, and drawing the string tight at different points to give it the required form.

Run the leg wires up through the leg, and wind with tow to the proper size. Push the wires through the body, and fasten them. If any special position is required for the tail, a wire may be run through the body into it; otherwise it may merely be pinned to the stand until dry. Having completed the wiring and stuffing, sew up the skin; bring the legs over the body, par-

allel to each other, and make the required bend
at the knees. Now mount your specimen in
such a manner as you may choose; put in the
eyes and set away to dry. There will be deficien-
cies, here and there, where the body does not
quite fill out the skin. These must be sup-
plied with chopped tow, before sewing up the
skin.

You can get the size and curves of the body
only by practice; but these few words on the
subject may be found of assistance to you;
remembering that all quadrupeds curve greatly
from the top of the hips to the tail.

If the animal is to be mounted with the mouth
open, place pieces of wood between the jaws,
and stuff out the lips in a natural manner until
dry, — when the props may be removed. A
tongue is made of cork or light wood, with two
wires secured to the back, by which it is afterwards fastened to the skull. Cover your artificial
tongue with wax, and place in position. The inside of the mouth and the gums must also be
neatly covered with wax. The whole now re-
quires to be painted with the color most resem-

bling nature, and when that is dry brushed over with a mixture of Damar varnish and oil of turpentine.

The best stand on which to mount mammals is an oval block of wood varying in thickness according to the size of the specimen. The name should be painted in black letters on the side of the block, and the whole varnished. Rocks, stump effects, etc., are made by bending pasteboard to the required shape, fastening to a standard, and stiffening with glue. Sand, smalts, etc., may then be dusted on. If there are several mammals mounted in the same case, a water-color background is very effective.

CHAPTER IV.

SKINNING AND MOUNTING FISHES AND REP-
TILES.

FISHES. — SKINNING.

THESE may be opened in two ways, according to the position in which the specimen is to be mounted. If the fish is to rest on its belly, an incision should be made the entire length of the ventral surface, from the gills to the end of the tail. If the fish is to rest on its side, the incision should be made on the side. Before proceeding farther cover the entire fish with tissue paper which will adhere with the use of thin gum.

Now, with the scalpel, handle carefully, detach the soft parts from the skin, cutting rib-bones with the scissors, until the back is reached. Cut through the fin-bones, and the body will be found quite loose. Detach the tail end, and re-

move all muscle from the remaining vertebræ.
Cut through the body at the base of the skull;
clean brain cavity thoroughly, and remove eyes.
This latter operation may require some assistance
from the scissors, on the outside. All muscles
about the eyes and skull should be carefully removed. When your skin is ready, poison it well
with the arsenic-alum powder.

MOUNTING.

The artificial body for your specimen may be
made of the same materials as used in stuffing
birds and mammals, of clay, plaster of Paris, or
the skin may be simply dried. A tow body may
be made and covered with a layer of clay, to give
it a smooth, even surface. You may form a
mould by pressing your specimen into damp
clay, allowing this to dry and then coating the
mould with colored varnish. When this is dry,
pour plaster of Paris of the consistency of cream
into the mould and let dry. The other side of
the fish must be treated in the same way, and the
two halves united by the solution of plaster.
When your body is ready, place it in the skin

and sew up. Place the specimen in the required position and fasten to a board by stout pins driven on each side. Spread the fins, tail, etc., by means of the wooden clamps already mentioned (Fig. 3, L), and set the specimen away to dry. A very convenient way of treating many specimens, especially hard-scaled fish, is to bring the sides of the opening together by a few stitches, and glue a strip of cloth the entire length of the incision. Before this is done, however, the end of the tail beyond the anus must be stuffed out with cotton. Take a few stitches through the gills to hold them down while drying. Now place a tin tunnel in the fish's mouth, and fill out the skin with fine sand. Place a wad of cotton in the throat, to keep the sand in; put the specimen in the desired position; remove the tissue paper with sponge and water; and set your specimen away for several weeks, to dry. When you are ready to mount your specimen, make several small holes in it, to let the sand out, and when quite empty fasten to a board; mount in a case, or in any way which your taste may suggest. It is sometimes desirable to retain

only one side of a specimen. That side should
be covered with tissue paper, as directed, and the
other side, soft parts, bone, etc., cut away.
Poison, place the skin on a board, and pin or
nail the edges fast, that it may not contract while
drying. Mount specimens with glass eyes, and
brush over with a coat of varnish. If spots, etc.,
fade, they must be touched up with paint.

REPTILES. — SKINNING.

Snakes, frogs, etc.. may be opened along the
belly, or they may be skinned through the mouth.
If the latter, open the mouth as wide as possible,
and with the scissors cut through the body and
first vertebra. Seize the stump with a pair of
forceps, and carefully push the muscles from the
skin, at the same time drawing the body out of
the mouth. This, of course, inverts the skin.
Poison thoroughly.

MOUNTING.

The best way to treat frogs is to fill out the
skin with sand, and when dry let the sand out of
it through pin holes. Put in eyes and varnish.

Snakes may be stuffed out with sand, or a body may be made. For the latter, take a piece of annealed wire, rather shorter than the specimen, wind with tow to the required size, and place in the skin. The wire enables you to give the specimen any position desired; while, if sand is used, the specimen must either lie coiled up or straight. If the mouth is to be kept open, a tongue may be made of fine wire and painted red.

CHAPTER V.

EGGS AND NESTS.

A FULL set of eggs is always desirable, **if they** can be obtained, but, as the old saying is, "A half-loaf is better than no bread." The contents may be removed by making a hole in the side of the egg with an egg drill, and sucking out the white and yolk with a glass blow-pipe. or by means of a little syringe with a bit of rubber tubing attached to the nozzle. If the young have already formed, a squarish-shaped hole may be made on one side, and the contents hooked out. The hole may be afterwards closed by pasting a bit of film or tissue paper over it. While drilling through the shell, the egg should be held over water, so that if dropped it may not be broken; or an arrangement made of wire resembling a pair of scissors, the ends terminating in a ring or oval, may be used. The ends are then covered with netting; thus forming a soft, yet strong, resting-place for the egg. (Fig. 5.)

The name of the specimen, together with size, date of collection and collector's name, should be written on the shell of each egg, and the entire hatch returned to the nest. It would be a good plan to give the eggs the same number as the parent bird, if this is obtained, together with a number

FIG. 5.

of their own. You can then note them in your ornithological ledger, or, if you choose, you can keep an oölogical ledger separate.

Nests should be preserved, if possible, attached to the branch on which they were found. This stem should be from three to six inches long, and be attached by its base to a block standard. Or, nests may be placed in little glass trays, made of pieces of window-glass held together by strips of paper glued over the edges. If the nest is not cared for, or cannot be obtained, the eggs may be placed on cotton, in little boxes, and arranged in the cabinet to suit the collector. A very good and safe way of transporting eggs, is to place them between two layers of cotton in the nest, which must be packed closely, but without pressure.

GUM PASTE.

Gum Arabic, 4 ounces.
Corrosive Sublimate, 2 grains.
White Sugar Candy, 2 ounces.
Mix.

Melt, and label "Gum Paste, for closing the holes drilled in eggs," etc.

CHAPTER VI.

SKELETONS.

DURING the busy collecting season, *rough* skeletons may be made by removing skin, viscera, and as much muscle as possible, covering the body with the arsenic-alum powder, and allowing it to dry, when the specimen may be wrapped in paper and laid away for future use. To prepare skeletons for the cabinet, remove as much of the fleshy parts as possible, and boil the bones until the remaining flesh is softened and can be easily removed. Then boil in water in which a piece of lime as large as a hen's egg has been dissolved. Remove, dry, and if necessary wire.

Another way recommended is to remove all the soft parts, and scald the hard parts in boiling water containing a few drops of hydrochloric acid. Leave the bones in this solution for ten minutes, wash, and boil in plain water until all the muscle, etc., is softened. Clean this away with a brush or by a stream of water. Boil in a strong solution of soda, wash with soap and water, and when perfectly clean, dehydrate with boiling alcohol (Junker). Skeletons should be mounted on wires fixed in a wooden standard painted black.

CHAPTER VII.

HUNTING AND HYGIENE.

To be a good collector, it is necessary to be something more than a good marksman. You must know at what time of day to go out to be most successful, and the localities where you are most likely to find the birds that you are looking for. In the field, you must be all eyes and ears. No thicket should be too dense, no tree too tall for your quick eye to penetrate its foliage ; no chirp or rustle too small or weak for your active ear to detect. In short, to be a good collector you must understand wood-craft. Sometimes a bird is seen just disappearing into the under-brush. A very good call, which seldom fails in bringing the bird from its retreat, is made by placing the back of the hand to the lips and sucking. By practice, this may be made to re-semble the cries of a wounded bird. Early

morning and just before sun-down have been
found to be the best hours for collecting, although
something may be done at any time of day.
During the noon hours, birds generally remain
hidden in the cool depths of the thickets and
woods. Birds are seldom found in the deep
forest; but, at the hours mentioned, trees and
bushes skirting roads, fields and meadows, will be
found teeming with life.

THE GUN.

The choice of a gun for collecting purposes is,
of course, optional with the reader; but a good
twelve or fourteen bore breech-loading shot gun
will give better satisfaction than any other,
and will be worth the price of the gun in time-
saving, when in the field. The pistol-guns, intro-
duced within the last few years, often prove of
great service in collecting small specimens.

LOADING.

Several sizes of shot should be taken into the
field, ranging from dust or mustard seed, as it is
sometimes called, to No. 6 or 8. For all small

birds up to the size of a robin, dust shot should be used ; and I have even killed grouse and plover with it, although, it must be confessed, at a very short range. The larger sizes are used for hawks and all large birds. With several sizes of shot, you can vary your loads according to the bird that you are pursuing. Here the breech-loading gun is vastly superior to the old-fashioned muzzle-loader, for it is but the work of an instant to change from large to small, or *vice versa.* Be careful not to load too heavily. Most of your birds will be killed within a few yards, and it is astonishing how little powder and how few shot will produce the desired effect.

CARRYING.

The most convenient and safe way to carry birds in the field, is in a common fish-creel ; or in a basket which I devised and have used for several years. It is simply a long, deep and narrow basket, carried on the back by straps which cross in front of the chest. At the back - of the basket, outside, is a netting for carrying paper, etc. ; and on either side a pocket or pouch of cloth for cotton, etc. For all l*.*ds under the size of a crow, this basket is very convenient.

Before going out provide yourself with a num-ber of sheets of stiff paper. As soon as a bird is shot, fill the mouth, anus, and shot-holes with cotton, and drop the bird head foremost, with bill pointing downwards, into a cornucopia of the paper, just the size of the bird's body, and fold the edges over the tail, taking care not to rumple or break the tail feathers. When birds are shot, they do not always die at once; but they may be put out of misery by placing the thumb under one wing, and the forefinger under the other, and squeezing. After a second or so, the bird will give a gasp and die.

This cannot be done in the case of large birds. To kill these, insert a thin knife-blade between the skull and last vertebra, cutting through the spinal cord; or break the back by pressing upon it with the knee.

TRANSPORTING.

Skins may be either packed in boxes, between layers of cotton, or they may be pushed head first into cylinders of stiff paper having a diameter equal to the largest part of the skin.

COLLECTING SUIT.

A serviceable and comfortable hunting-suit may be made from any good strong stuff, such as corduroy, etc. The pants should be made rather loose, and have the seams firmly sewed. The coat should be a mere succession of pockets, and of course very loose. A soft, broad-brimmed felt hat, and a pair of broad-soled, low-heeled shoes, for ordinary wear; or, for shooting where the country is wet and boggy, a pair of high top boots may be substituted. This will be found to be the easiest, most durable and least expensive outfit that can be made.

EATING.

Do not start out in the morning without having first partaken of a lunch, however slight, as a preventive, if nothing more; for tramping on an empty stomach will almost always upset one for the whole day.

AT HOME.

As soon as you return from your day's tramp, a good "wash-up" and a change of clothes will rest you far more than sitting down. Especially, if your feet are wet, lose no time in changing socks, and all other garments that are in the least damp. By doing this, you will save yourself many a severe cold, and perhaps a fit of sickness.

POISONING.

In case of poisoning with the arsenic, while preparing your skins, the advice of Dr. Coues, in his "Field Ornithology," covers the whole treatment. "Avoid," he says, "all mechanical irritation of the inflamed parts, touch the parts that have ulcerated with a stick of lunar caustic; take a dose of salts; use syrup of iodide of iron, or tincture of chloride of iron, say thirty drops in a wine-glass of water, thrice a day; rest at first; exercise gradually as soon as you can bear it; and skin no birds till you have completely recovered." If these do not cure, medical advice should be procured.

HINTS ON

WRITING AND SPEECH-MAKING

BY

THOMAS WENTWORTH HIGGINSON,

AUTHOR OF

"Young Folks' History of the United States," "Young
Folks' American Explorers," "Malbone,"

"Outdoor Papers," "Oldport Days," Army Life in
A Black Regiment," "Atlantic Essays," etc.

Price, 50 cents.

"So delicate and yet so strong in the style; So apt, yet so
abundant his illustrations; So fascinating the easy, polished
leisurely diction, that the literary enjoyment cannot be im-
paired. He has all the charms of Montaigne without his
egotism."— *Minneapolis Press.*

The following from the author's Preface shows the purpose
of this volume: —

"The first of these two chapters appeared long since in the
Atlantic Monthly, and was afterwards included in the author's
volume entitled 'Atlantic Essays.' Teachers have several
times urged that it should be reprinted as a little manual of
literary composition; and I have indeed seen it used for that
purpose in a college class-room. Now that similar suggestions are beginning to come in respecting the other brief essay,
'Hints on Speech-making.' it has seemed well to present the
two together in a small volume, The last-named paper ap
peared first in 'Harper's Magazine' and it is here reprinted
with the consent of the publishers."

HANDBOOK

OF

LIGHT GYMNASTICS.

By LUCY B. HUNT,

Instructor in Gymnastics at Smith (Female) College, Northampton, Mass.

Illustrated. Cloth, 60 cts.

" This manual is admirably adapted for the use of teachers and pupils in public and private schools, and in seminaries and colleges, as well as a guide to health-giving exercises in the homes, especially for girls. Well-arranged series of exercises are given in free gymnastics, wand exercises, ring exercises, dumb-bells, procession, mutual-help exercises, bean-bags, marching, and a practical chapter on dress suitable for gymnastic exercises to be taken in. All these exercises have been carefully selected and thoroughly tested, and can be safely practised by any person in ordinary health." — *Journal of Education.*

"Taking the system of Dr. Dio Lewis as a foundation, Miss Hunt has, during her experience as a teacher, taken from, added to, and altered various exercises, until the course, so to speak, has assumed the order now presented in her little book, a course which, if carefully followed, will make the maidens of America better fitted to become its mothers." — *N. Y. Mail and Express.*

" It is designed as a guide to teachers of girls, but it will be found a use also to such as wish to practice the exercises at home." — *N. Y. World.*

" A volume so very diminutive that one can hardly realize that it contains nearly all that one needs for the teaching or practice of light gymnastics, and even more than Dr. Lewis's clever and amusing volume." — *Budget.*

" This work has many advantages. It is inexpensive, it is convenient, it is condensed, it is clear. It is careful to avoid any strained, unnatural, or ungraceful positions, it does not attempt to make a gospel of gymnastics, as some fanatics have done." — *N. Y. Christian Advocate.*

" A useful little manual, by a teacher of much experience, who embodies in this little work the best results of her knowledg and practice of the modern system of gymnastic exercises fo girls' schools and colleges; and also for use at home." — *Jerus June.*

EXERCISES
FOR THE IMPROVEMENT OF THE SENSES
FOR YOUNG CHILDREN.

By HORACE GRANT,
Author of "Arithmetic for Young Children."

Edited by Willard Small.
Cloth. Price, 50 cents.

For the purpose of producing instruction and amusement
for young children, too young to read or write, this little work
has been prepared. The special object is to excite little chil
dren to examine surrounding objects correctly, so that valuable
knowledge may be acquired, while the attention, memory,
judgment, and invention are duly exercised.

In exercises such as those which compose this book, the most
favorable circumstances may be seized as they arise, and will
therefore produce an extraordinary effect. Wherever we are,
in a room, garden, field, or road, in the morning or evening,
winter or summer, action or rest, something interesting may
be extracted; for at the moment when the attention is warmly
excited, an event may be turned to the best account. The val
uable habits acquired by means of familiar objects and petty
events may gradually be extended to the most important
subjects.

ARITHMETIC FOR YOUNG CHILDREN

Being a series of Exercises exemplifying the manner in which
Arithmetic should be taught to young children

By HORACE GRANT.
American Edition, Edited by WILLARD SMALL.
Price 50 cents.

"Consists of a series of exercises illustrative of the manner
in which the first steps in numbers should be taught to young
children. We pronounce it *first-rate*. The primary teacher
will find it a great aid in her work. It is rational and consis-
tent. The variety of style and method used lend fresh interest at every step."—*Educational Weekly.*

"The forms of expression used and the copiousness of illustrations are very far in advance of the common style of
teaching this science to young children.

"It is thoroughly rational and prepares the way for a more
systematic study of numbers as the child becomes more ma-
ture. The young pupil is taught to think and speak in numbers in the first stage and subsequently unites with it the art
of writing numbers. It is correct in theory and apt for
practice."—*N. E. Journal of Education.*

THE DEBATER'S HANDBOOK.

Including a Debate on the Character of Julius Cæsar, adapted for J. SHERIDAN KNOWLES, author of "The Hunchback," "William Tell," and other famous plays, designed for practical Exercises in Declamation, and as a Model for Debating Clubs; also for Classes in Public and Private Schools, with directions for forming and conducting debating clubs and societies, rules of debate, list of subjects and references, etc.

Cloth, 50 cents. Boards, 50 cents. Paper, 30 cents.

The contents of this volume, beside the celebrated Debate, contains "Rules for Debate," with directions and suggestions, 100 selected questions for debate, with reference lists, etc. Of the "Julius Cæsar" Debate it has been said, "It is doubtful whether the English language can furnish any matter more appropriate for the application of the principles of elocution, or better adapted to use as a *practical exercise* in declamation."

"The author well deserved the statue which stands in front of the Massachusetts State House."

A FEW THOUGHTS FOR A YOUNG MAN.

By HORACE MANN.

Cloth, 50 cents.

In 1849, Horace Mann, one of the most skilful and sympathetic educators known to this country, delivered a lecture before the Mercantile Association, Boston, choosing for his theme: "Thoughts for a Young Man." So kind, fatherly, wise, and sagacious were the counsels to young men embodied in this lecture, that its publication was immediately sought for; and the work enjoyed a wide popular circulation. Sundry editions have been redemanded; and it now appears from the publishing house of Lee & Shepard, Boston. The book is earnestly commended to young men for its sober recall from the deceptive glitter of material things, its lofty inculcations and its wholesome precepts. The author well deserved the statue which stands in front of the Massachusetts State House. — *Newark Advertiser.*

Punctuation and Other Typographical Matters.

For the use of Printers, Authors, Teachers, and Scholars. By MARSHALL T. BIGELOW, Corrector at the University Press, Cambridge. Small 4to. CLOTH, 50 CENTS.

Lenox Library, New York, Aug. 19, 1881.

DEAR MR. BIGELOW, — I sent for your "Punctuation and other Typographical Matters" (having long groaned over bad pointing in authors and printers), and was glad to find an excellent manual which will contribute to the comfort of many. I cordially recommend it to all authors, printers, and men of letters.

Faithfully yours,

Allibone's Dictionary of Authors.   I. AUSTIN ALLIBONE.

"Mr. Bigelow's book is a practical treatment of the subject, and enlarges the reading public's obligations to him." — *Atlantic Monthly.*

"It is intended for the use of authors and teachers, while business men who have occasion to print circulars, advertisements, etc., can hardly afford to be without a copy of it for reference." — *Schenectady Daily Union.*

Mistakes in Writing English, and How to Avoid Them.

For the Use of all who Teach, Write, or Speak the Language. By MARSHALL T. BIGELOW, author of "Punctuation and other Typographical Matters." CLOTH, 50 CENTS.

"This is an admirable little work; the more admirable for the use of busy people, because it is little, since it is also clear and comprehensive. The errors pointed out are those to which nearly all writers are liable. . . . We commend it as the most convenient little manual of which we have knowledge." — *Christian Herald.*

"This is a valuable little volume. It is not a grammar, with rules and definitions; but it takes up words and parts of speech, and shows, generally by example, their correct use. It is arranged systematically, and is adapted to the use of the home and the school." — *The Current.*

"The matter is well arranged, and the points upon which instruction is desired can be readily found." — *Christian Union.*

"This is a useful book. A careful study of the several chapters would be of great advantage to all who have to do much or little speaking or writing." — *Gospel Banner.*

CAMPBELL'S
HANDBOOK OF ENGLISH SYNONYMS
WITH AN APPENDIX,
SHOWING
THE CORRECT USES OF PREPOSITIONS.

160 pages. Neat cloth binding, 50 cts.

This compact little volume contains about 40,000 synonymous words, printed in clear, distinct type.

It is a work which will substantially aid speakers, writers, teachers and students — in fact all who would gain a more copious vocabulary and increase their power of expression.

It includes the really important matter of the more bulky volumes which are commonly sold for two dollars or more.

A great choice of words is here placed at the service of the writer and the speaker.

The Appendix, containing "Prepositions Compared and Discriminated," and "A List showing what Prepositions to use after certain Words," is a trustworthy guide in a great number of cases of doubtful usage. A writer's knowledge of English idiom and his style are best shown by his use of these little hinges of the language.

LEE AND SHEPARD'S POPULAR HANDBOOKS

Price, each, in cloth, 50 cents, except when other price is given.

Exercises for the Improvement of the Senses. For Young Children. By HORACE GRANT, author of "Arithmetic for Young Children." Edited by WILLARD SMALL.

Hints on Language in connection with Sight-Reading and Writing in Primary and Intermediate Schools. By S. ARTHUR BENT, A.M., Superintendent of Public Schools, Clinton, Mass.

The Hunter's Handbook. Containing lists of provisions and camp paraphernalia, and hints on the fire, cooking utensils, etc.; with approved receipts for camp-cookery. By "AN OLD HUNTER."

Universal Phonography; or, Shorthand by the "Allen Method." A self-instructor. By G. G. ALLEN.

Hints and Helps for those who Write, Print, or Read. By B. DREW, proof-reader.

Pronouncing Handbook of Three Thousand Words often Mispronounced. By R. SOULE and L. J. CAMPBELL.

Short Studies of American Authors. By THOMAS WENTWORTH HIGGINSON.

The Stars and the Earth; or, Thoughts upon Space, Time, and Eternity. With an introduction by THOMAS HILL, D.D., LL.D.

Handbook of the Earth. Natural Methods in Geography. By LOUISA PARSONS HOPKINS, teacher of Normal Methods in the Swain Free School, New Bedford.

Natural-History Plays. Dialogues and Recitations for School Exhibitions. By LOUISA P. HOPKINS.

The Telephone. An account of the phenomena of Electricity, Magnetism, and Sound, with directions for making a speaking-telephone. By Professor A. E. DOLBEAR.

Lessons on Manners. By EDITH E. WIGGIN.

Water Analysis. A Handbook for Water-Drinkers. By G. L. AUSTIN, M.D.

Handbook of Light Gymnastics. By LUCY B. HUNT, instructor in gymnastics at Smith (female) College, Northampton, Mass.

The Parlor Gardener. A Treatise on the House-Culture of Ornamental Plants. By CORNELIA J. RANDOLPH. With illustrations.

Sold by all booksellers, and sent by mail, postpaid, on receipt of price.

LEE AND SHEPARD Publishers Boston

LEE AND SHEPARD'S POPULAR HANDBOOKS

Price, each, in cloth, 50 cents, except when other price is given.

Warrington's Manual. A Manual for the Information of Officers and Members of Legislatures, Conventions, Societies, etc., in the practical governing and membership of all such bodies, according to the Parliamentary Law and Practice in the United States. By W. S. ROBINSON (*Warrington*).

Practical Boat-Sailing. By DOUGLAS FRAZAR. Classic size, $1.00. With numerous diagrams and illustrations.

Handbook of Wood Engraving. With practical instructions in the art, for persons wishing to learn without an instructor. By WILLIAM A. EMERSON. Illustrated. Price $1.00.

Five-Minute Recitations. Selected and arranged by WALTER K FOBES.

Five-Minute Declamations. Selected and arranged by WALTER K. FOBES.

Five-Minute Readings for Young Ladies. Selected and adapted by WALTER K. FOBES.

Educational Psychology. A Treatise for Parents and Educators. By LOUISE PARSONS HOPKINS, Supervisor in Boston Public Schools.

The Nation in a Nutshell. A Rapid Outline of American History. By GEORGE MAKEPEACE TOWLE.

English Synonymes Discriminated. By RICHARD WHATELY, D.D., Archbishop of Dublin. A new edition.

Hints on Writing and Speech-making. By THOMAS WENTWORTH HIGGINSON.

Arithmetic for Young Children. Being a series of Exercises exemplifying the manner in which Arithmetic should be taught to young children. By HORACE GRANT. American Edition. Edited by WILLARD SMALL.

Bridge Disasters in America. The Cause and the Remedy. By Prof. GEORGE L. VOSE.

A Few Thoughts for a Young Man. By HORACE MANN. A new Edition.

Handbook of Debate. The Character of Julius Cæsar. Adapted from J. SHERIDAN KNOWLES. Arranged for Practice in Speaking, for Debating Clubs, and Classes in Public and Private Schools.

Sold by all booksellers, and sent by mail, postpaid, on receipt of price

LEE AND SHEPARD Publishers Boston

www.ingramcontent.com/pod-product-compliance
Lightning Source LLC
Chambersburg PA
CBHW021525090426
42739CB00007B/778